AF382954

As Editor, I want to
thank my wife
Esnur, my parents
Rico and Ursula
Micheroli-Blum, my
mentors and friends
Giorgio Tamborrini,
Adrian Ciurea,
Diego Kyburz and
Thomas Brack.

I dedicate this book
to Can.

Raphael Micheroli

THE RHEUMATIC HAND

**EDITORS AND
AUTHORS**
Raphael Micheroli, MD
Fellow sonographer
Kantonsspital Glarus
Burgstrasse 99
8750 Glarus, Switzerland

Giorgio Tamborrini, MD, KD
Head of departement
Ultrasound SGUM/Sonar/
EFSUMB/Eular/QIR
Rheumatology FMH
Ultrasound Center
c/o Bethesda Hospital
4020 Basel, Switzerland

Diego Kyburz, MD, Prof.,
Chief of departement
Rheumatology
University Hospital of Basel
4020 Basel, Switzerland

Glarus, 2015

**PRODUCTION AND
PUBLISHER**
BoD – Books on Demand,
Norderstedt
ISBN 978-3-7347-7882-7

**EDITORIAL DESIGN
AND PHOTOS**
Janine Wiget, Graphic
Design & Illustration

CONTENTS

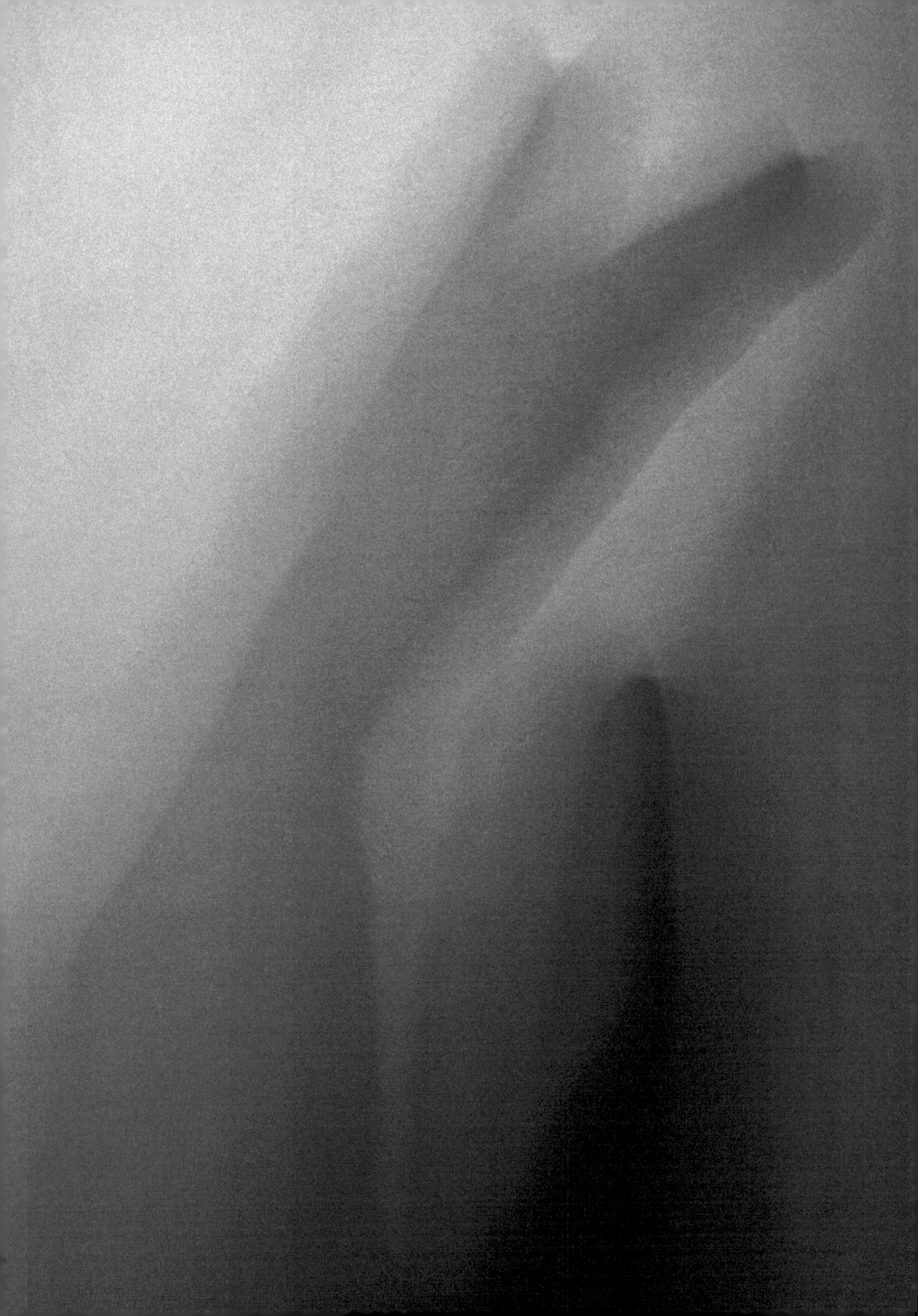

INTRODUCTION

Many rheumatic diseases show changes that are visible in the hands. Fundamental information on these diseases is revealed by the pattern of distribution in the relevant joints, soft-tissue changes, skin manifestations, neurological and vascular symptoms, and clinical findings. Imaging and lab results provide diagnostic support.

In this book, these common diseases are presented in terms of their clinical expressions in the hands: osteoarthritis, rheumatoid arthritis, gout, calcium pyrophosphate dihydrate deposition disease, psoriatic arthritis, reactive arthritis, systemic sclerosis and dermatomyositis/polymyositis. Furthermore, we discuss the pathological findings in the hands as a result of diabetic cheiroarthropathy, endocarditis, secondary hypertrophic osteoarthropathy and chronic regional pain syndrome.

R. MICHEROLI
G. TAMBORRINI
D. KYBURZ

RHEUMATIC-
DISEASES

OSTEOARTHRITIS (OA)

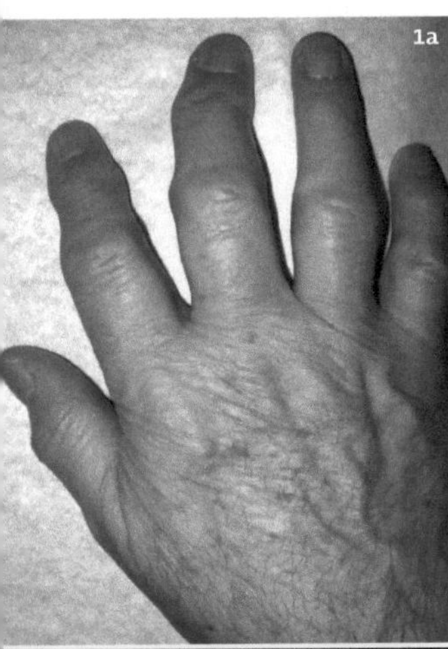

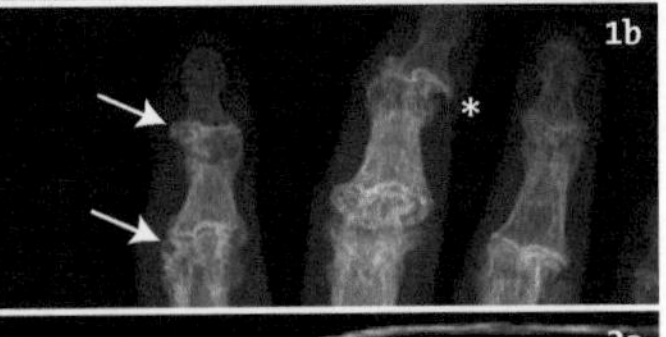

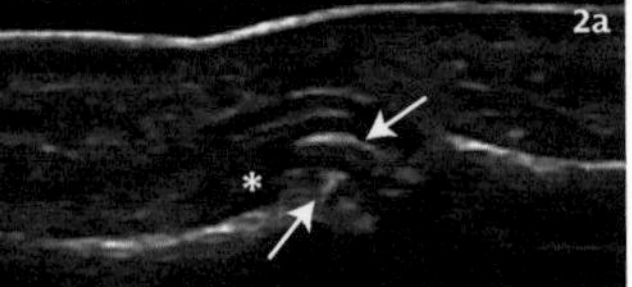

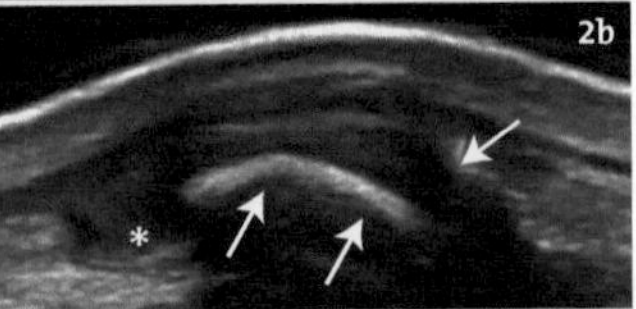

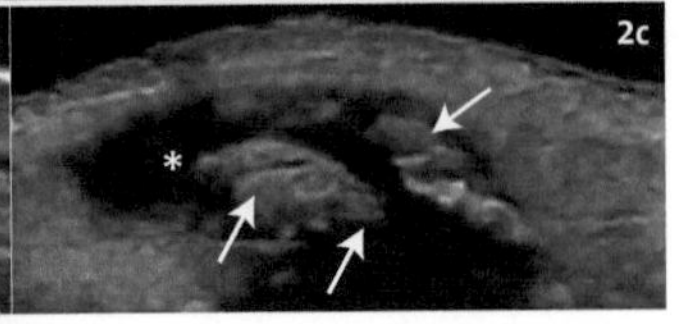

1a Clinical image showing swelling and deformity of the distal and proximal interphalangeal joints

1b X-ray image with joint space narrowing (arrowhead), subchondral cyst (arrow), osteophytes (star) and deformities of the distal and proximal interphalangeal joints

2a-c Ultrasound of the volar longitudinal DIP joint (at the top) and of the longitudinal PIP joint (at the bottom, 2D and 3D) with osteophytes (arrows) and joint effusion (stars)

EPIDEMIOLOGY

Osteoarthritis is the most common joint disease. The incidence of coxarthrosis (hip OA) is 40-80/100,000 inhabitants/year, that of gonarthrosis (knee osteoarthritis) is 160-240/100,000 inhabitants/year, and that of finger osteoarthritis is 100/100,000 inhabitants/year. In general, women are affected approximately 1.7 times more frequently. The prevalence in 34-year-olds is as high as 17% but increases to over 90% in people aged 65 and over.

ETIOLOGY AND PATHOGENESIS

The causes of primary idiopathic osteoarthritis are unclear. However, genetic predisposition (association with HLA-A1,-B8), endocrine factors and metabolic disorders seem to have an influence.

Risk factors for secondary osteoarthritis include age, gender (women aged 55 and over are especially affected), joint trauma, obesity, dietary factors, biomechanical malalignment, deformity, inflammatory diseases and diseases of the nervous system. Chondrozytic dysfunction, which is associated with an inadequate synthesis capacitiy, leads to an overall degradation of the cartilage and to a secondary loss of tissue surrounding the joint.

HAND MANIFESTATIONS

Osteoarthritis of the finger joints is usually a primary, polyarticular osteoarthritis with an affinity for the distal interphalangeal (DIP) joints. Heberden's osteoarthritis refers to the involvement of the DIP joints. Bouchard's osteoarthritis is a proximal interphalangeal (PIP) joint osteoarthritis that occurs simultaneously with Heberden's osteoarthritis in every third patient. In contrast to Heberden's osteoarthritis, Bouchard's osteoarthritis often has no knots. Rhizarthrosis (first carpometacarpal joint osteoarthritis) leads to more severe functional deficits than Heberden's or Bouchard's osteoarthritis. Erosive (destructive) osteoarthritis may occur as an inflammatory (activated) osteoarthritis variant. Osteoarthritis of the fingers is not always painful; however, it can lead to cosmetic disfigurement.

FURTHER CLINICAL MANIFESTATIONS

Patients often present with the following symptoms: impact pain, fatigue pain and stress pain. Over time, constant pain, night pain and muscle pain (localized, "joint-associated") can occur. In very severe cases of osteoarthritis, thickening and bony deformity, instability and muscular atrophy as well as stiffening of the deformity has been observed. The classic signs of inflammation occur inn activated osteoarthritis and include pain, hyperthermia, swelling, redness and impairment.

DIAGNOSIS

Conventional x-rays show typical changes that include asymmetric ("eccentric") joint space narrowing, subchondral sclerosis, the building up of osteophytes with cysts and, in severe cases, deformity and erosions. Small osteophytes or joint effusion can be observed using ultrasound (US). In primary osteoarthritis, laboratory findings alone are not of diagnostic value. Joint aspiration is not inflammatory (less than 2000 cells/µL to 200-500 cells/µL, predominantly mononuclear inflammatory cells) and may include hydroxyapatite crystals.

THERAPY

A cure for osteoarthritis does not exist. The main goals of therapy are pain relief, maintenance of function and control of the degenerative processes.

The basic treatments are NSAIDs and chondroprotective agents. When activated osteoarthritis is present, intra-articular glucocorticoids can be used. In many cases, physical therapy or orthopedic measures should be considered. The ultima ratio is joint replacement surgery.

RHEUMATOID ARTHRITIS (RA)

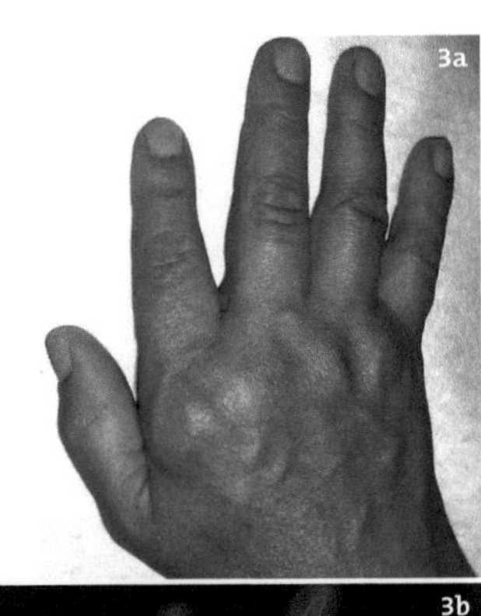

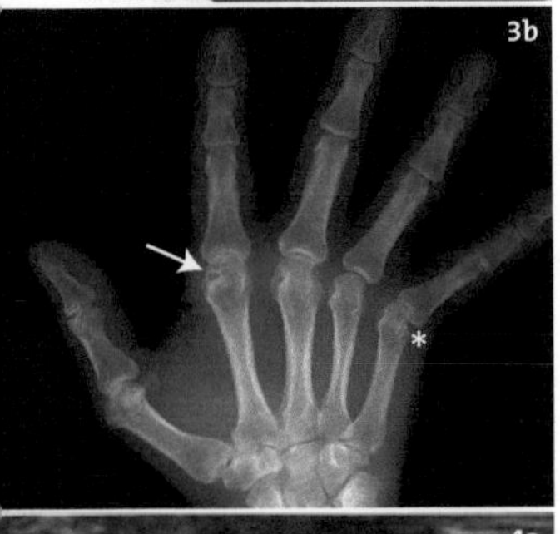

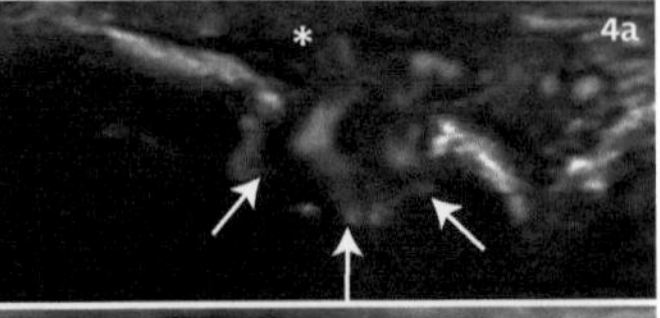

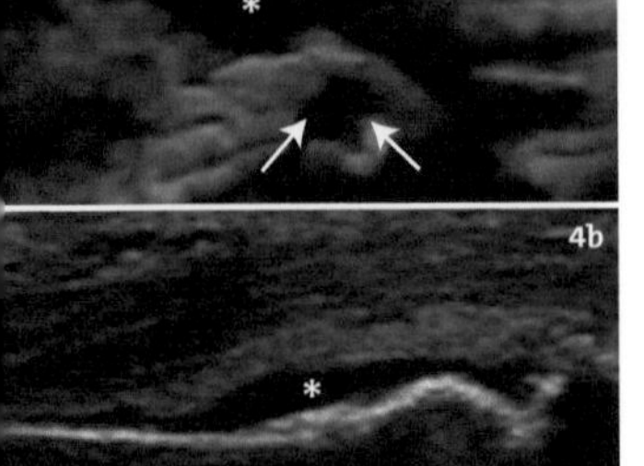

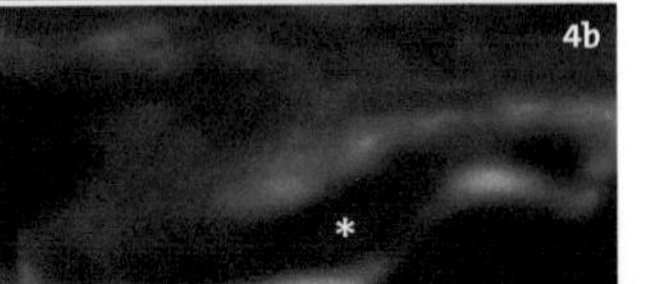

3a Clinical picture showing ulnar deviation of the fingers and joint swelling of the MCP joints

3b X-ray image showing joint space narrowing in the MCP joints II and III, subluxation of the MCP joint V (star) and a cyst in the head of metacarpal II (arrow)

4 a/b Ultrasound of MCP joint II dorsal longitudinal (**4a**) and volar (**4b**) with synovitis (stars), abnormal power Doppler signal (=active inflammation) and erosion (arrows)

EPIDEMIOLOGY

The most common inflammatory rheumatic disease affects approximately 1% of the population. The peak age of onset is between 40 and 60 years; however rheumatoid arthritis (RA) may occur at any age. Women are affected three times as commonly as men.

ETIOLOGY AND PATHOGENESIS

It is believed that the cause is multifactorial, with genetic factors and environmental influences playing major roles. RA is strongly associated with major histocompatibility complex class II alleles and human leucocyte antigen (HLA)-DR4. Non-genetic factors may include infectious agents, hormones and smoking. An unknown trigger starts the inflammation in the synovium of a joint. Different cells are activated: immune cells activated by cytokines (TNF-alpha, IL-1/18/6) maintain the inflammation; B cells produce autoantibodies (rheumatoid factor, anti-CCP Ab); osteoclasts cause bone loss; and fibroblasts produce proteases, which break down collagen and cartilage. The result of the cellular activation is cartilage loss and bone resorption with the formation of erosions. The inflammation may spread to capsules, tendons and ligaments, and muscles may atrophy. Over time, the joints lose their function due to the impairments.

HAND MANIFESTATIONS

MCP, PIP and wrist joints typically have a

symmetrical synovitis. Additionally, tenosynovitis frequently occurs. Other characteristics include atrophy of the interossei muscles, swan neck deformities, boutonniere deformities, tendon ruptures, 90°/90° deformity of the thumb, volar subluxation of the wrist with step building, contractures of the finger bender, inability to stretch and to make a fist, ulnar deviation of the fingers, dislocation of the distal radius and distal ulna, and rheumatoid nodules. Vasculitic lesions, such as ulcers, may occur in long-standing RA. Neurologically, carpal tunnel syndrome can develop and is often due to a tenosynovitis in the carpal tunnel or radiocarpal synovitis.

FURTHER CLINICAL MANIFESTATIONS

Symptoms of RA include joint pain (resting pain, night pain, improvement with movement, constant pain), stiffness of the joints (morning stiffness >one hour) and loss of strength. Patients experience general symptoms such as fatigue, adynamia, anorexia, weight loss and low-grade fever. Large joints and the upper cervical spine can also be affected. The skin can have ulcers and rheumatoid nodules, and the heart may be affected with pericarditis and rheumatoid nodules in the myocardium. Pleurisy, pleural effusion, interstitial fibrosis or intrapulmonary rheumatoid nodules can be observed in the lungs. Sicca symptoms, scleritis in the context of systemic vasculitis and neurological manifestations (cervical myelopathy with cervical involvement) may also occur.

DIAGNOSIS

The diagnosis can be made clinically. RA has typical radiographic findings of periarticular soft tissue swelling, periarticular osteopenia, joint space narrowing, bone erosions (over time), joint destruction and subluxation with ankylosis. Ultrasound can detect inflammatory changes (e.g., synovitis, tenosynovitis) that cannot be diagnosed with a standard X-ray image. The presentation of cartilage or bone changes, such as cartilage thinning or erosions, is observed earlier with US. Rheumatoid factor and anti-CCP antibodies serve as diagnostic markers. If both are positive, the sensitivity and specificity of the tests increase. The joint aspiration is inflammatory (>2000 cells, predominantly polynuclear inflammatory cells).

THERAPY

Treatment goals include the suppression of the inflammatory response and the prevention of bone destruction. The drugs used, based on the risk profile, include corticosteroids, cDMARDs, and bDMARDs (e.g., methotrexate, leflunomide, salazopyrin, and biologics). An early initiation of the basic therapy is crucial to slowing the progression of the disease. Non-drug treatments include physical therapy and occupational therapy.

GOUT (URATE ARTHROPATHY)

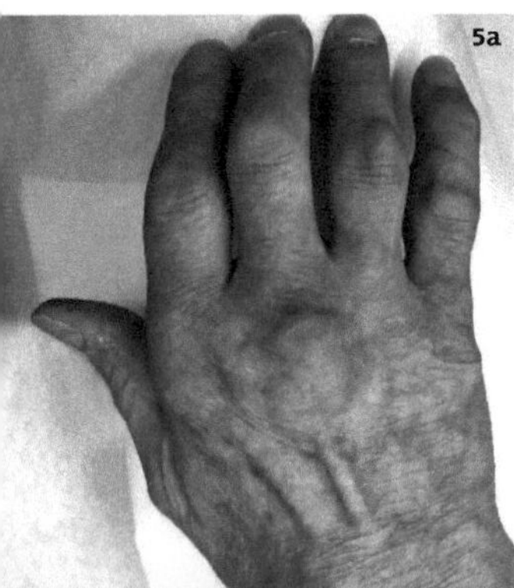

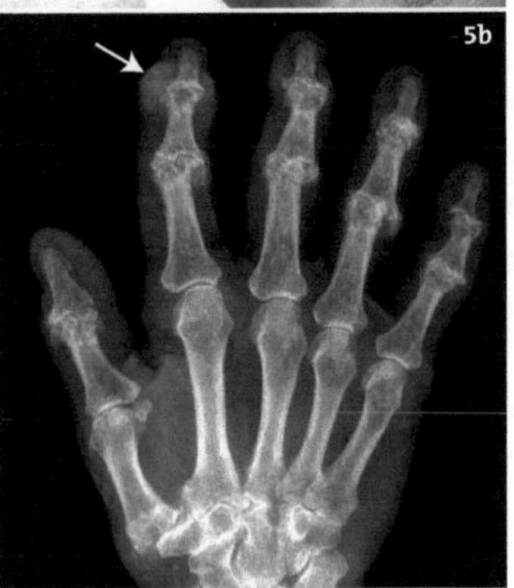

5a Clinical picture of poly-articular gout with soft tissue tophi over the PIP joint 4 and the metacarpal head III and radiocarpal synovitis.

5b Destruction in almost all PIP and DIP joints with soft part tophi radial of the DIP joint II (arrow)

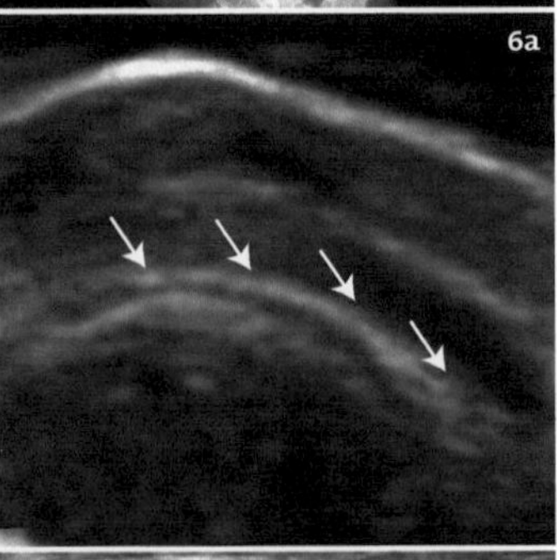

6a/b Dorsal longitudinal ultrasound of MCP joints with an erosion of the metacarpal head (star) and a double contour (arrows) over the cartilage of the metacarpal head

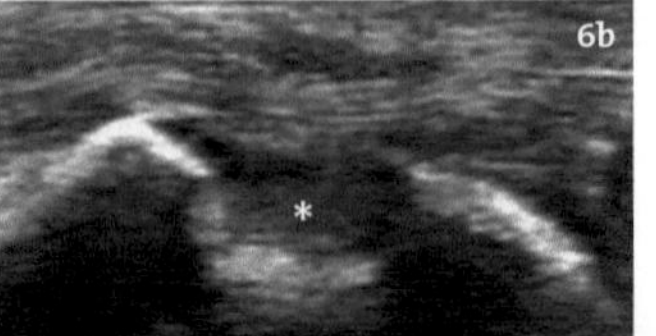

EPIDEMIOLOGY

Gout is a common crystal arthropathy that occurs in 2.8% of men and 0.4% of women aged 30-59 years. Gout concerns men 10 times more commonly than women. Up to 20% of the population is affected by hyperuricemia. The prevalence of gout depends on the age and standard of living. The main age of manifestation in men is between the fourth and fifth decades of life, whereas in women, gout is most common after menopause, usually in the sixth decade of life.

ETIOLOGY AND PATHOGENESIS

Risk factors include metabolic syndromes (diabetes mellitus, hypertension, hyperlipidemia and obesity), increased alcohol consumption and renal failure.

Gout is an acute, sometimes chronic crystal-induced arthritis resulting from hyperuricemia. The primary hyperuricemia occurs due to a decreased excretion of uric acid by the kidneys. The secondary form is caused by a reduced renal excretion of uric acid due to kidney disease or medications (e.g., diuretics). Hyperuricemia can also occur through the increased production of uric acid associated with hematological diseases (i.e., myeloproliferative disorders, leukemia, and lymphoma) or due to chemotherapy for malignant tumors. Uric acid accumulates, exceeds the solubility product and deposits in bradytrophic tissues.

Crystal phagocytosis activates a cytokine cascade and other inflammatory mediators and thus triggers an acute attack of gout.

HAND MANIFESTATIONS

Acute gouty arthritis in a metacarpophalangeal joint is called chiragra. Periarthritis with soft tissue edema and redness of the skin occur concomitantly with acute gout. In chronic gout, polyarticular involvement can occur and may mimic rheumatoid arthritis or other crystal arthropathies such as calcium pyrophosphate dihydrate deposition disease.

Usually, the symptoms begin at night or early in the morning with intense swelling, redness and severe tenderness. An attack typically has a duration of hours to several days; however, without treatment, it can last weeks. Soft tissue tophi may occur in the hand.

FURTHER CLINICAL MANIFESTATIONS

Gout manifests as acute monoarthritis initially in 90% of the cases. In 50% of cases, it is localized to the first metatarsophalangeal joint (podagra). Other sites, in descending order of frequency, include ankles, wrists, knees and joints of the upper extremities. Concomitantly, periarthritis and a strong collateral edema occur.

Tophi occur on the feet, ears, bursae, tendons, subcutaneous tissues, and as bone tophi in the joints or even in the spine.

DIAGNOSIS

An X-ray image can indicate large tophi and erosions. Ultrasound shows typical crystal deposits, which lie on the hyaline cartilage (double contour sign), tophi, erosions or hyperechoic synovitis. Laboratory tests show an inflammatory response with increased blood sedimentation rate, elevated C-reactive protein and leukocytosis. In acute gouty arthritis, hyperuricemia often cannot be found. Evidence for the gouty arthritis includes the presence of urate crystals in joint aspiration or specific ultrasound findings. The joint aspiration is inflammatory (more than 2000 cells with predominantly polymorphonuclear inflammatory cells).

THERAPY

The treatment of acute gout includes intra-articular steroids, NSAIDs (beginning with high doses, immediate and long-acting pharmaceutical technology), systemic steroids or peroral colchicine. Drinking sufficient fluids, normalization of weight, reduction of alcohol consumption (especially beer), avoidance of medications that increase serum uric acid, and treatment with uricostatics (inhibition of synthesis) and uricosuric agents (increased excretion of uric acid) lead to a lower serum uric acid.

CALCIUM PYRO-PHOSPHATE DIHYDRATE DEPOSITION DISEASE (CPPD)

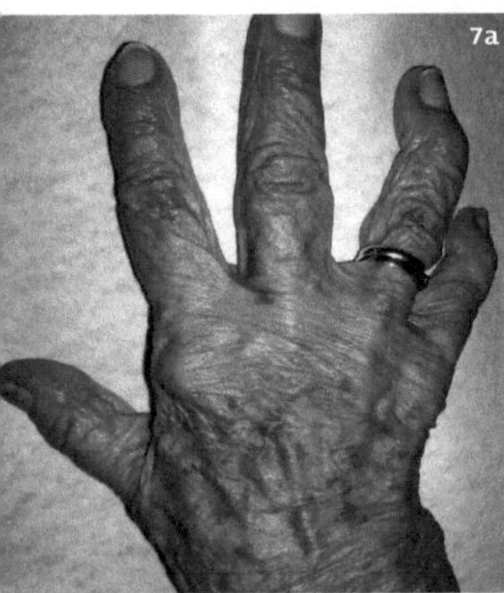

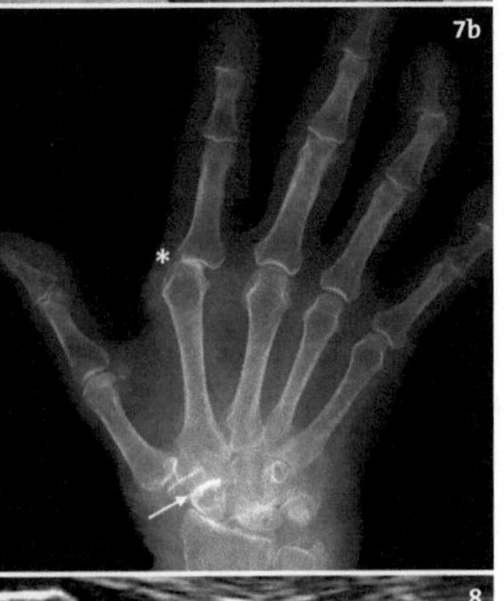

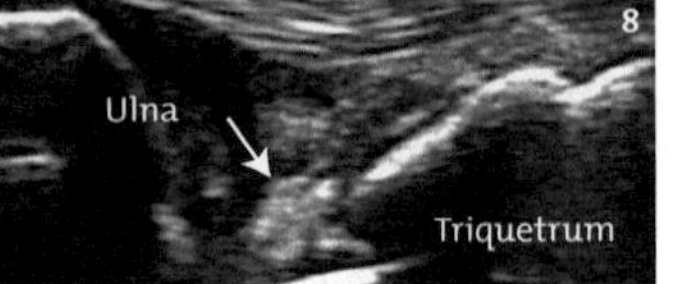

7a Clinical picture with dislocated head of the os metacarpale of the index finger

7b X-ray image with narrowing of the joint space and subluxation in the MCP joint II (star) and STT osteoarthritis (arrow)

8 Ultrasound of the dorsal wrist carpoulnar, longitudinal, with calcification in the TFCC (arrow)

EPIDEMIOLOGY

The disease is characterised by deposits of calcium pyrophosphate in hyaline and fibrous cartilage, synovial membranes, tendons and ligaments and affects approximately 1% of the population. The typical age of onset is between 60-80 years. Women are 5 times more frequently affected than men.

ETIOLOGY AND PATHOGENESIS

Calcium pyrophosphate dihydrate deposition disease (CPPD, pseudogout) is mostly idiopathic, and the main risk factor is age. Hemochromatosis, hyperparathyroidism, hypophosphatemia and hypomagnesemia may promote calcium pyrophosphate dihydrate deposits. The pathogenesis consists of two phases. The first phase occurs because of a malfunction of the enzyme pyrophosphatase with consecutive formation and deposition of crystals in the cartilage matrix. In the second phase, these crystals cause an inflammatory and/or catabolic-destructive reaction.

HAND MANIFESTATIONS

CPPD manifests in the hand in different ways. A clinically silent form occurs that involves the formation of a secondary osteoarthritis, typically in the STT joint (scapho-trapezio-trapezoideal joint), in the radiocarpal joint and the MCP-II and -III joints (large hook-like osteophytes). An acute monoarthritis (pseudogout with periarthritis) may occur. In addition, a chronic, rarely erosive polyarthritis (not to be confused with rheumatoid arthritis), can develop. Tenosynovitis occurs frequently.

FURTHER CLINICAL MANIFESTATIONS

Chondrocalcinosis is usually discovered accidentally during radiological examinations. A trauma-induced attack of pseudogout occurs as monoarthritis with swelling, redness and hyperthermia.

The knee joints are most commonly affected. Deposits in the soft tissue around the dens axis lead to a pseudo-meningitis ("crowned dens syndrome"). Bursae, tendons and ligaments may calcify and cause pain. CPPD is manifested in 50% of the cases as secondary osteoarthritis. In the severe, destructive form, it leads to erosion and destruction or deformity of the knee, hip and shoulder joints.

DIAGNOSIS

In addition to the above-described osteoarthritis in the STT joint, MCP joint II and III, X-ray and ultrasound can show calcification of the triangular fibrocartilaginous complex (TFCC), of the ligaments of the proximal carpal row, or inside the cartilage. Usually, the laboratory tests show increased inflammation parameters. If there are other underlying diseases (such as endocrinopathies and other conditions) some parameters can change; for example, calcium and parathyroid hormone can be increased and ferritin, magnesium, phosphate or TSH can be lowered. In addition to the pathognomonic radiographic signs, either the microscopic evidence of calcium pyrophosphate crystals from the joint aspiration or the typical findings using high-resolution ultrasound are essential for a reliable diagnosis.

THERAPY

In the latent form, no therapy is necessary. NSAID, steroids, colchicine, and intra-articular steroids are used for acute therapy.

There is currently no effective drug treatment available to prevent the progression of the joint destruction. Therefore, therapy is mainly directed at treating the symptoms. Magnesium, which plays a role in the activity of pyrophosphatase and the solubility of CPPD crystals, may be beneficial. Methotrexate can be tried if all other therapies fail.

PSORIATIC ARTHRITIS (PSA)

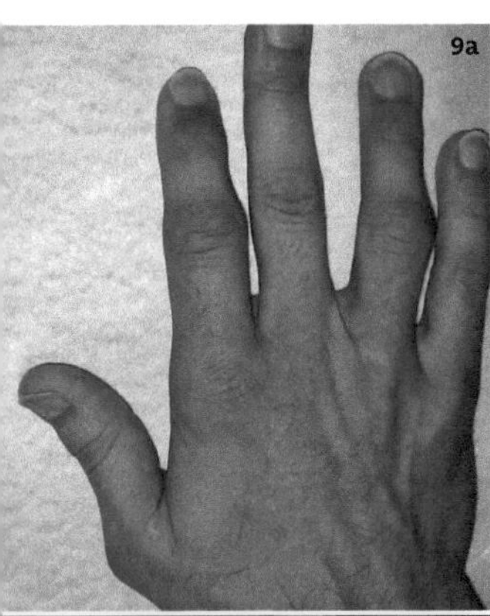

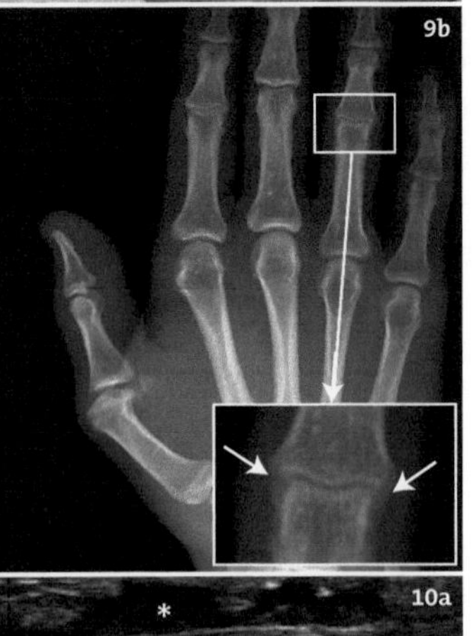

9a Clinical picture with swelling of the PIP joints II and IV

9b X-ray image with bony proliferations in the PIP joints II and IV (arrows)

10a Ultrasound of the PIP joint, palmar longitudinal (at the top) with tenosynovitis of the flexor tendons (star) and synovitis in the PIP joint (arrow);

10b DIP joint (star), dorsal, coronal and 3D view with increased vascularization at the enthesis and nail bed (arrow head)

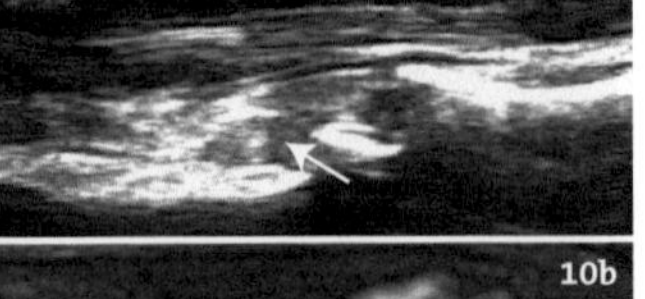

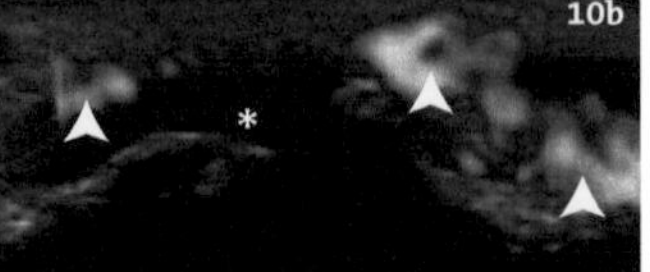

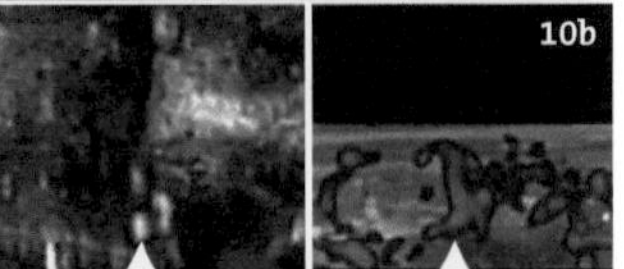

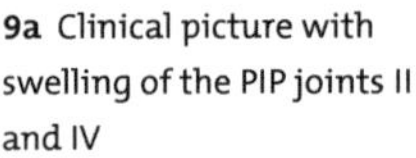

EPIDEMIOLOGY

The skin disease psoriasis occurs in 1-2% of the population. Psoriatic arthritis belongs to the spondylarthritides and affects 0.1-0.5% of the population. Five to ten percent of the dermatologically affected patients have psoriatic arthritis. Men and women are equally affected. Of patients with exclusive joint and axial skeleton involvement, men are more often affected. The main age of onset for psoriatic arthritis is between the 3rd and 4th decades of life, but it can occur at any age.

ETIOLOGY AND PATHOGENESIS

In 40% of cases, the family history is positive for skin psoriasis or psoriatic arthritis. Genetically, an association with HLA-B27 is detectable. The possible roles of a vitamin D3 metabolism disorder, overexpression of proto-oncogenes and T-cell activation by keratinocytes have also been discussed.

HAND MANIFESTATIONS

Psoriatic arthritis has different manifestations in the hand and can occur as a monoarthritis (especially PIP, DIP, wrist), as an asymmetric oligoarthritis (fewer than 4 joints affected) or as a polyarthritis. A transversal distribution is typical and usually involves the DIP joints with nail changes or involves all of the joints of one finger (MCP, PIP and DIP joints). Psoriatic arthritis rarely results in a mutilating arthritis. Nail psoriasis typically includes the following changes: stippling, crumb nails, white spots, lateral grooves, sub-ungual keratosis of the nail bed, onycholysis and oil spots. In psoriatic arthritis with nail involvement, an enthesitis of the tendons in the distal phalanx and a DIP joint arthritis is often present. Psoriatic dactylitis corresponds to a lymphedema-like swelling of a finger (sausage fingers) with underlying tenosynovitis, soft tissue edema and synovitis.

FURTHER CLINICAL MANIFESTATIONS

Further Clinical Manifestations The psoriasis is mainly found in the stretch-side knee and elbow region, the sacral region, the scalp and the external ear canal.

Throughout the investigation, the entire body should be examined for the possibility of an inverse variant. The spine, especially the sacroiliac joint, can be affected by inflammation in the spondyloarthritis. In addition, in the musculoskeletal system, an inflammatory enthesopathy (=enthesitis) at the heel or knee, a tenosynovitis, or arthritis of the joints of the anterior chest wall can develop. Uveitis and inflammatory bowel changes may occur.

DIAGNOSIS

X-ray and ultrasound show a mixed picture of erosions or joint destruction (pencil in cup changes), bony proliferations, ankylosis, enthesitis, arthritis and soft-tissue swelling. Spinal radiograph frequently shows unilateral sacroiliitis and parasyndesmophytes (paradiscal intervertebral bone formation). Ultrasound is more sensitive than X-ray and clinical examination in the assessment of the disease activity (arthritis, enthesitis). Laboratory tests can show an increase in BSR and CRP, the HLA-B27 may be positive, and the joint aspirate is inflammatory.

THERAPY

Treatment consists of NSAIDs, DMARDs (methotrexate, leflunomide, sulfasalazine) and TNF-alpha inhibitors. Individual joints can also be treated with intra-articular steroid injection. Treatment accompanied by consultation with a dermatologist is important.

REACTIVE
ARTHRITIS (REA)

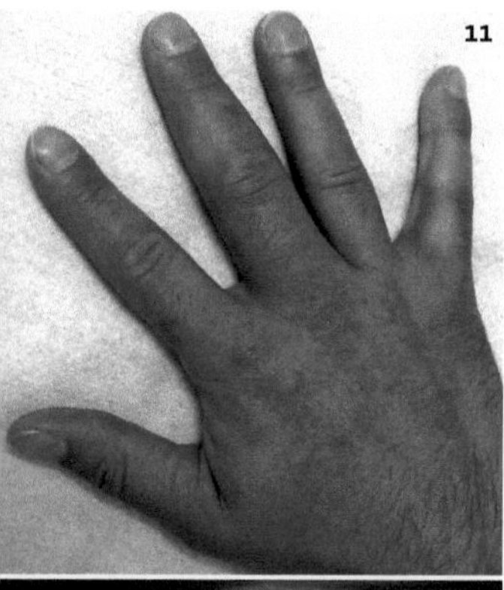

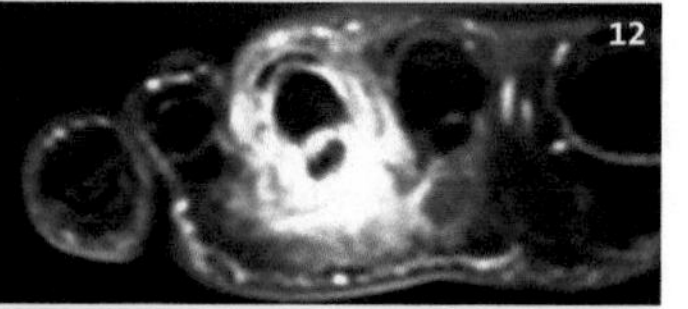

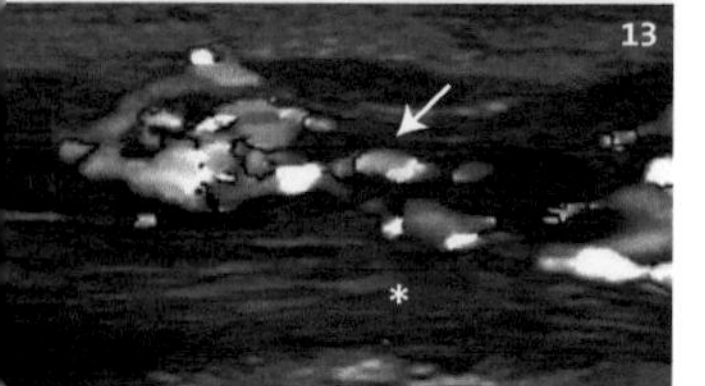

11 Clinical features
of dactylitis of the middle
finger

12 Coronary MRI with
hyperintense swelling
around the middle finger

13 Ultrasound of the
flexor tendon (star) of the
middle finger with evidence
of tenosynovitis with an
abnormal power Doppler
signal (arrow)

EPIDEMIOLOGY

Thirty to forty persons/100,000/year show
an asymmetrical, non-infectious oligoarthritis
after clinically and/or microbiologically doc-
umented intestinal or genitourinary infection
(infection is not always detectable). The typical age of onset is between 20 and 40 years.
The gender distribution of women:men is 1:1
in postenteritic infection, in posturethritic
infections, the distribution is 1:20.

ETIOLOGY AND PATHOGENESIS

A reactive arthritis occurs because of an
immunological reaction in the synovial membrane after contact with bacterial antigens.
This form of reaction genetically determined
(HLA-B27) to some extent.

Reactive spondyloarthritides (HLA-B27
associated) can be triggered by the following
infections: Chlamydia trachomatis/pneumo-
niae, Salmonella, Yersinia enterocolitica, Shi-
gella flexneri, Campylobacter jejuni/fetus or
Clostridium difficile. There is no association
between HLA-B27 and Borrelia burgdorferi,
Staphylococcus aureus, Neisseria gonorrhoe-
ae, beta-hemolytic streptococci, Mycobacte-
rium tuberculosis or Gardnerella vaginalis.

HAND MANIFESTATIONS

Similar to the other spondyloarthritides
(e.g., ankylosing spondylitis, psoriatic arthritis, undifferentiated spondyloarthropathy
and enteropathic spondyloarthritis), there

may be a dactylitis. Often, there is a mono- or asymmetric oligoarthritis. Fingers and wrists can be affected, and tenosynovitis or enthesitis can occur. On the skin, keratoderma blenorrhagicum (pustulating erythema on the palms and soles, and later a hyperkeratotic, scaly, oozing lesion) may be found.

FURTHER CLINICAL MANIFESTATIONS

This arthritis is manifested as asymmetrical mono/oligoarthritis of large joints, predominantly of the lower extremities. There can be inflammation of periarticular structures such as peritendinitis, tenosynovitis and enthesitis.

The skin shows the following changes: mucosal ulcerations (oral, genital), balanitis, erythema nodosum, and keratoderma blenorrhagicum (foot sole). Conjunctivitis, keratitis or anterior uveitis may occur. If monoarthritis, urethritis and conjunctivitis are present, a reactive arthritis is very likely.

DIAGNOSIS

A history of previous infections, together with the clinical findings, helps to make a diagnosis. Only in chronic cases does X-ray show a mixed picture of erosions, proliferation and ankylosis. Laboratory analyses detect increased signs of inflammation. The PCR for chlamydia can be positive in the often highly inflammatory synovial fluid but is more frequently identified in the urine. Stool cultures allow for the detection of enteric infections. The HLA-B27 can be positive and is considered a risk factor for chronic cases.

THERAPY

Selective treatment with antibiotics can be applied in certain cases if an infective agent that has led to a reactive arthritis persists. The arthritis is treated with NSAIDs, intra-articular glucocorticoids and, in chronic cases, with sulfasalazine or methotrexate.

SYSTEMIC SCLEROSIS (SSC)

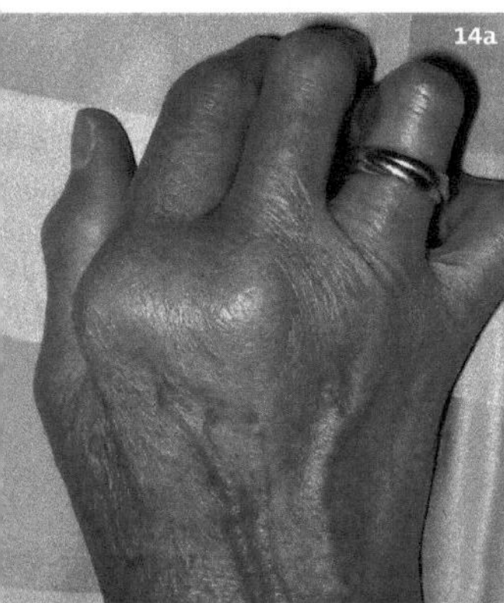

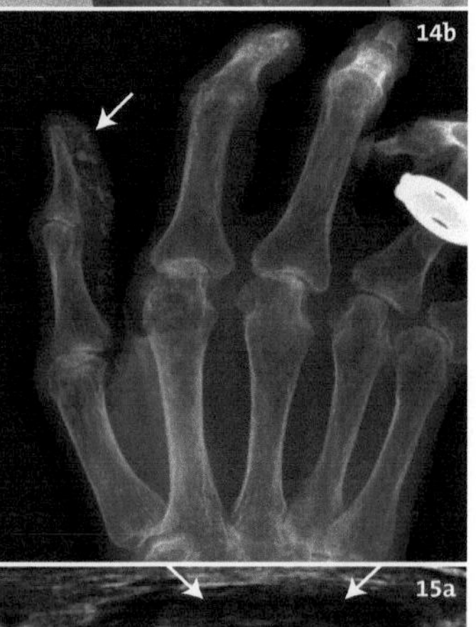

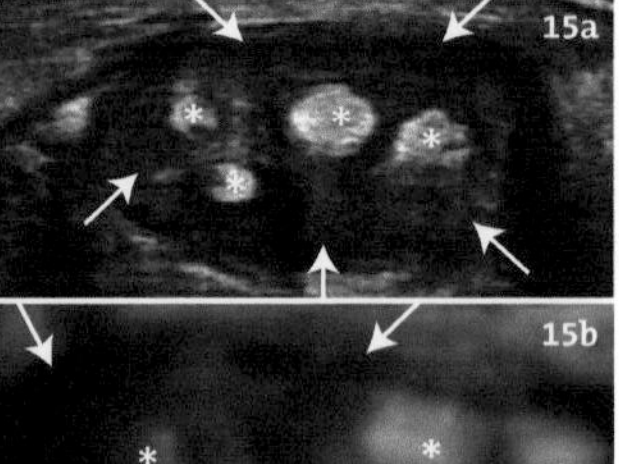

14a Clinical picture with flexion contractures, Madonna fingers, swelling over the MCP joints and sclerotic skin

14b X-ray image showing contractures, post-inflammatory changes with joint space narrowing in the MCP joint II and III and dystrophic calcifications in the soft tissues of the thumb (arrow)

15a/b Ultrasound image (2D and 3D) of the forth extensor compartement with a chronic isoechoic tenosynovitis (arrows) around the tendons (stars)

EPIDEMIOLOGY

Systemic sclerosis (SSc) is a rare autoimmune disease. The incidence is 4-14/1 million inhabitants/year, and the prevalence is 100-140/1 million inhabitants. Women are affected three times more often than men. The disease mainly occurs between the ages of 35 and 65.

ETIOLOGY AND PATHOGENESIS

An association with different HLA alleles and chromosomal abnormalities seems to play a role. In addition, exogenous factors (chemical and fumed silica) have been discussed as causes. The pathogenic basis is an autoimmune vascular and connective tissue disease, and inflammatory as well as fibrotic changes are involved. The clinical picture is caused by vascular changes such as a change in the vascular morphology of capillaries and a reduction, false activation and immune system dysfunction, and the pathological accumulation of extracellular matrix.

HAND MANIFESTATIONS

Often, arthralgia and arthritis of the small finger joints (usually the MCP joints) and tenosynovitis are present. Raynaud's phenomenon is an early and common initial symptom that occurs in 90 % of SSc cases. The condition is cold induced and leads to vasoconstriction with a subsequent blanching, bluish discoloration due to cyanosis, and finally to a reactive vasodilatation with redness, swelling, throbbing and burning pain. In the early stages of the disease in the hands, edema of the fingers (puffy fingers) occurs. The actual sclerosis, with induration, atrophy and increased fibrosis of the skin, follows later. It is possible that skin depigmentation, hyperpigmentation or telangiectasia occur, and the soft tissue may have dystrophic calcifications. Diffuse thickening of the fingers with livid discoloration is called sclerodactylitis. A claw-like fixed flexion of the fingers ("Madonna finger") and vascular necrosis (ulceration) in the fingertips may occur over time.

FURTHER CLINICAL MANIFESTATIONS

In the face, SSc can cause microstomia (narrowing of the mouth), atrophy of the lips, skin puckering around the mouth ("tobacco pouch mouth") and tongue band reduction and sclerosing. The skin of the whole body can become scleritized. Joint symptoms can include pain, swelling (synovitis), crepitus and stiffness. Internal organs are very commonly affected. Esophageal stiffness with atony, loss of peristalsis, gastroesophageal reflux and esophagitis occur early and in 90% of cases. Pulmonary involvement occurs in 60% of patients and is usually the main reason for premature death. Cardiac involvement may be primary (pulmonary arterial hypertension) or secondary caused by lung disease. In the kidneys, systemic sclerosis manifests in the form of an acute renal failure (renal crisis, mainly as a result of high-dose steroid use). Other affected organs include the gastrointestinal tract, skeletal muscle, blood vessels, liver, pancreas, and rarely the central nervous system and eyes.

DIAGNOSIS

In 2013, the ACR and the EULAR described skin thickening of the fingers extending proximal to the metacarpophalangeal joints as sufficient for the patient to be classified as having SSc. If such thickening is not present, then 7 additional items apply, with varying weights for each: skin thickening of the fingers, fingertip lesions, telangiectasia, abnormal nail fold capillaries, interstitial lung disease or pulmonary arterial hypertension, Raynaud's phenomenon, and SSc-related autoantibodies (antinuclear antibody and antibodies against various defined nuclear antigens). Depending on the organ system involvement, further tests are performed, including chest X-ray, echocardiography, right heart catheterization, pulmonary function tests, and esophageal manometry.

THERAPY

SSc can only be treated symptomatically because the disease process cannot be stopped. The therapy depends on the affected organs (e.g., calcium antagonists in Raynaud, vasodilators, proton pump inhibitors, immunosuppressants, anti-fibrotic drugs, etc.).

DERMATOMYOSITIS POLYMYOSITIS (DM/PM)

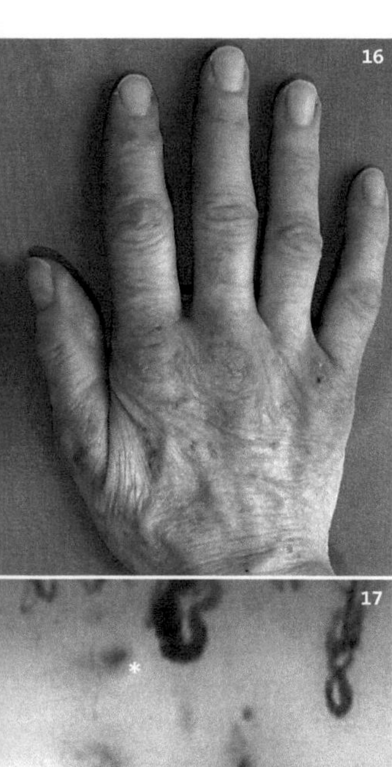

16 Clinical features of Gottron papules over the MCP joints

17 Capillary microscopy: arrow: microhemorrhages, star: dilated capillaries

EPIDEMIOLOGY

Dermatomyositis (DM) and polymyositis (PM) are rare diseases.

The prevalence is approximately 60/1 million inhabitants, the incidence 2-10/1 million inhabitants/year. Skin manifestations (dermatomyositis) are detectable in 20-30% of cases. Women are two to three times more frequently affected than men. The disease has its main manifestation between 40 and 50 years of age.

ETIOLOGY AND PATHOGENESIS

The cause is unknown. Different HLA alleles and other genetic factors are suspected as potential causes. Environmental factors, infections, drugs and vaccines also seem to have an influence, but the causality has not been proved.

A type-4 immune reaction in which antibodies bind to endothelial antigens and lead to complement activation is held responsible for muscle damage, which results in C5b-9 membrane attack complex deposits. These deposits lead to capillary necrosis and perivascular inflammatory infiltrates, i.e., CD4 T-lymphocytes, B-lymphocytes, and plasmacytoid dendritic cells with swelling of the endothelium and injury.

It is assumed that the skin changes occur through antibody-mediated cytotoxicity (antibody dependent cellular cytotoxicity, type-2 immune response).

Arthralgias and non-erosive arthritis occur, especially of the MCP joints. Pathognomonic lesions on the hand are called Gottron papules and are confluent, erythematous to purple papules located dorsally to the metacarpophalangeal and the interphalangeal joints. In the area of the nail fold, these papules can lead to painful, hyperkeratotic thickening (Keinig sign) when pressure is applied. Periungual telangiectasia and splinter hemorrhages can also occur. Under capillary microscopy, giant capillaries, micro-hemorrhages and disturbed capillary architecture can be identified. 'Mechanic's hands' (radial hyperkeratosis with scaling and fissures stressed) are observed in the context of antisynthetase syndrome (e.g., with anti-Jo-1 antibodies and pulmonary fibrosis).

FURTHER CLINICAL MANIFESTATIONS

In addition to nonspecific, general symptoms such as fatigue, fever and weight loss, the onset is heralded by a weakness of proximal muscles. The symptoms occur gradually and increase over time. In severe cases, the pharynx and/or respiratory muscles are also affected.

Other skin manifestations include a heliotrope (purple-colored) rash and an erythema induced by UV radiation (V sign, shawl sign). Especially in cases of juvenile DM we see calcinosis cutis. Pulmonary involvement in the form of interstitial pneumonitis, as well as a cardiac involvement, are not uncommon. Joint pain and non-erosive arthritis may occur. DM often occurs together with malignant diseases. A cancer screening must be discussed with each patient with DM.

DIAGNOSIS

History and symptoms are indicative. Magnetic resonance imaging (MRI) may show inflammatory changes in the muscles, and the electromyogram may have an abnormal pattern. Muscle biopsy with subsequent histological examination is the gold standard for confirming the diagnosis. Laboratory tests show increased muscle enzymes (CK, myoglobin, LDH, GOT and GPT). The following autoantibodies are increased: ANA (80%) and myositis-specific antibodies (e.g., anti-Jo-1, anti-Ku and anti-Mi2); the latter are only detectable in 30% of cases.

THERAPY

The foundations of the treatment of DM and PM are corticosteroids (prednisone) and immunosuppressants (e.g., methotrexate, azathioprine and cyclosporine).

SYSTEMIC LUPUS ERYTHEMATOSUS (SLE)

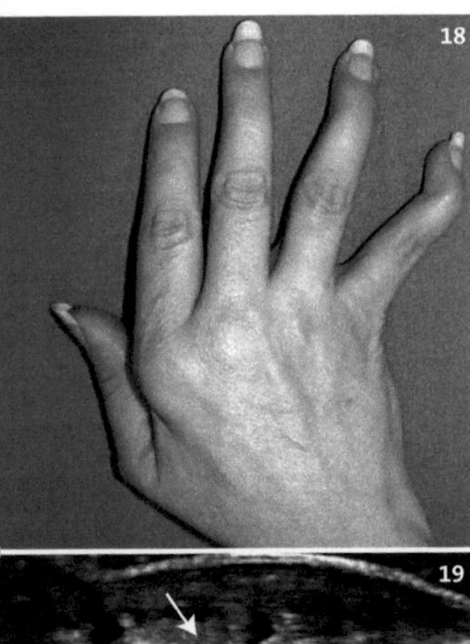

18 Clinical picture of Jaccoud arthritis

19 Ultrasound of a finger flexor tendon (star) with evidence of exudative and proliferative synovitis (arrow)

EPIDEMIOLOGY

The epidemiology of this autoimmune disease has strong regional differences. The incidence is 25-27 cases/100,000 inhabitants/year in Europe. The prevalence averages 25-27/100,000 inhabitants in Europe, 49/100,000 in Asia and 200/100,000 in Africa. Women develop the disease in 80-90% of cases, and the third decade of life is when the disease occurs most frequently.

ETIOLOGY AND PATHOGENESIS

The influence of hormonal factors (estrogen) and various environmental factors such as UV radiation, bacterial infections, viruses and drugs have been discussed. Etiologically, a genetic predisposition (HLA allele variations) has been considered.

SLE involves a disturbed immune balance in which B cells are hyper-reactive and T cells are attenuated, as found in all autoimmune diseases. B cells produce antibodies against the patient's own body, and deposits of immune complexes are found in organs (e.g., in the kidneys) and in the skin.

HAND MANIFESTATIONS

Symmetrical migratory polyarthritis or polyarthralgia are frequent. As with other connective tissue diseases, Raynaud's phenomenon and paronychial changes occur. The Jaccoud arthropathy, which is rare and occurs in fewer than 10% of patients, affects tendons and/or muscles and may cause the development of swan and/or ulnar deviation. Joints can dislocate. Bone erosions usually do not occur. In patients with acrocyanosis, there may be flat, blue-red hyperkeratotic nodes (Chilblain lupus). Erythema on the fingers is frequently observed dorsally between the joints.

FURTHER CLINICAL MANIFESTATIONS

Lupus erythematosus is one of the most multifaceted diseases. Although there many patients who suffer from the syndrome, only a few exhibit the whole disease spectrum.

Over 90 % of patients complain about general symptoms such as malaise, fatigue and intolerance. Polyarticular, symmetrical and migratory arthralgias occur with equal frequency. Myalgias are observed in 70 % of patients. The internal organs are also commonly affected. Pulmonary involvement in the form of pleurisy and cardiac involvement in the form of pericarditis occur in 60 % of cases. Neuropsychiatric lupus mostly manifests as headaches, intermittent cognitive impairment and depression. A diffuse proliferative glomerulonephritis is the most common form of lupus nephritis. Skin involvement occurs in 85 % of patients and consists of butterfly erythema in the face, alopecia (hair loss), intraoral/intranasal ulcers, photosensitivity, and discoid lesions (circular lesions with raised, erythematous borders and central hyperkeratosis).

DIAGNOSIS

The classification criteria can help with the diagnosis and include the following: butterfly erythema, discoid skin lesions, photosensitivity, ulcers, arthritis, serositis (pleurisy, pericarditis), renal disease, neuropsychiatric disorders (seizures or psychosis), hematologic disease (cytopenias), and immunological findings (anti-dsDNA Ab, anti-Sm Ab, anti-phospholipid antibodies and antinuclear antibodies).

The disease can be classified as SLE if four or more criteria are met. Diagnostic imaging may be helpful (e.g., transthoracic echocardiography for detecting cardiac involvement, MRI of the brain, or EEG for the detection of CNS involvement).

Early diagnosis of the disease is very difficult because SLE can masquerade for years oligosymptomatically, e.g., as undifferentiated arthritis.

THERAPY

Hydroxychloroquine and steroids are the mainstays in the treatment of SLE, especially when considering the possible need for rapid control of disease activity.

For the treatment of the organ manifestations, belimumab, mycophenolate mofetil, cyclophosphamide and B-cell depletion are used. In cases of predominantly joint involvement, methotrexate is used. Lupus patients should consistently protect themselves against the sun.

NON-
RHEUMATIC-
DISEASES

DIABETIC CHEIROARTH-ROPATHY (DIABETIC HAND SYNDROME)

EPIDEMIOLOGY

The prevalence in diabetes is between 8-58% (more common in type I diabetes than in type II). The disease usually occurs late but can also occur at an early stage of diabetes.

ETIOLOGY AND PATHOGENESIS

The risk of suffering from diabetic cheiroarthropathy increases with higher HbA1c values and disease duration of diabetes. Advanced age and smoking are also risk factors.

Deposits of abnormal collagen in the connective tissue appear to be responsible for the stiffness of the joints and the skin lesions. Enzymatic and non-enzymatic glycation of collagen, abnormal crosslinking of collagen and increased collagen hydration contribute to the pathogenesis. Microangiopathy and neuropathy promote contractures via fibrosis and misuse.

HAND MANIFESTATIONS

In diabetics, in addition to carpal tunnel syndrome and tenosynovitis, diabetic hand syndrome can manifest (cheiroarthropathy). This condition leads to a diffuse thickening of the soft tissues of the hand (fascia, tendons, ligaments and capsules), which limits the mobility, in particular the extension of the fingers. The joints themselves are not affected.

Diabetic cheiroarthropathy is usually painless. Clinically, the so-called "prayer sign" can be identified. A positive prayer sign means the patient is unable to touch the palms together as if assuming a prayer position because the fingers cannot straighten completely. However, this phenomenon can also occur in other diseases such as rheumatoid arthritis due to arthritis in the PIP joints and tenosynovitis due to other causes.

FURTHER CLINICAL MANIFESTATIONS

Diabetic hand syndrome occurs, as the name suggests, as part of diabetes. The other clinical manifestations of diabetic complications include the following: general non-specific symptoms, symptoms due to hyperglycemia and glucosuria with diuresis, symptoms due to disturbances of electrolyte and fluid balance, impotence, amenorrhea, macro-and microangiopathy, diabetic nephropathy, diabetic retinopathy, diabetic neuropathy, diabetic foot syndrome, diabetic cardiomyopathy, lipid metabolism disorders, fatty liver and hyporeninemic hypoaldosteronism.

DIAGNOSIS

The "prayers sign" and the "table top test" (the ability to brace the palm on the surface of a table) are helpful for diagnosis. Flexion and extension can be assessed using goniometry, and ultrasound may detect thickening of the tendon sheath and the skin.

THERAPY

Diabetes should be optimally treated. Physiotherapy can lead to an improvement in hand function. Smoking should be avoided.

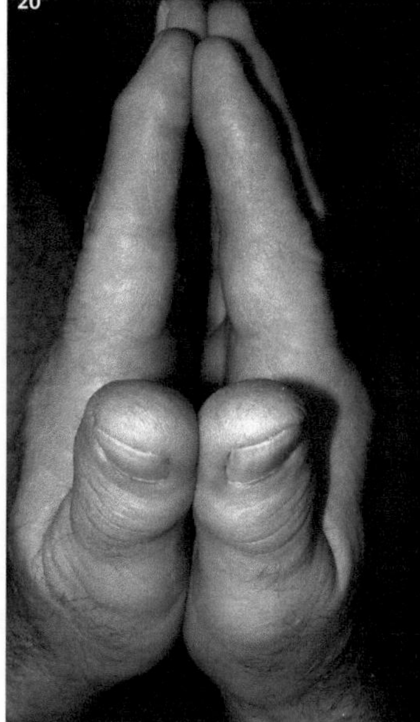

20 Clinical picture of the "prayers sign"

ENDOCARDITIS

EPIDEMIOLOGY

In Western Europe, infectious (bacterial) endocarditis occurs approximately 3/100,000 people/year. The women:men ratio is 3:2 to 9:1. The people most often affected are below the age of 60. The mitral and/or aortic valves are most commonly affected by the inflammatory.

ETIOLOGY AND PATHOGENESIS

The pathogens involved are 40-50% streptococci, 30-40% staphylococci, 10% enterococci and rarely Coxiella burnetii, Chlamydia, Mycoplasma, Legionella and pathogens of the HACEK group. Often, the blood culture remains negative. A pre-existing defect of the heart valves predisposes patients to endocarditis. Platelet-fibrin thrombi deposit on the area of endocardial lesions, and these thrombotic deposits are an ideal place for the colonization of bacteria. This colonization results in the local destruction of the valves and myocardial damage, embolization of vegetations in the periphery (tissue infarction, septic metastases), immune complex deposition and tissue destruction (glomerulonephritis, Osler's nodules).

HAND MANIFESTATIONS

Splinter hemorrhages are linear subungual hemorrhages that resemble wood chips under the nail. These hemorrhages are 1-3 mm long, red and change color from brown to black after a few days. Splinter hemorrhages can also be seen after trauma or antiphospholipid syndrome (primary or secondary in systemic lupus erythematosus) but seldom in rheumatoid arthritis, thromboangiitis obliterans, left atrial myxoma or atrial thrombosis. Osler's nodes are small, swollen and red areas with a bright center on the fingertip. These nodes may disappear after a few hours but usually remain visible for several days. Janeway lesions occur on the palms or soles of the feet and are small, erythematous or hemorrhagic spots. Watch-crystal nails and drumstick fingers are seen in some cases of endocarditis.

FURTHER CLINICAL MANIFESTATIONS

Endocarditis is associated in with fever in 90% of cases. General symptoms such as weakness, anorexia, weight loss, sweating, arthralgia and septic arthritis are also observed. The cardiac symptoms vary depending on which cardiac valve is infected. Over time, a myocardial abscess may occur.

The bacterial micro-emboli may lead to embolic encephalitis and can also settle in the retina. Renal involvement results in hematuria and proteinuria.

DIAGNOSIS

The history is revealing, and the presence of intravenous drug use, diagnostic or therapeutic interventions in patients with heart defects and related factors are important to note. The Duke criteria support the diagnosis if the following are present: 2 major criteria, 1 major criterion and 3 minor criteria, or five minor criteria. The major criteria are positive blood cultures with typical microorganisms for infective endocarditis from two separate blood cultures and evidence of endocardial involvement by echocardiogram. The minor criteria are predisposing heart conditions or intravenous drug use; fever >38°C, vascular findings (arterial emboli, septic pulmonary infarcts, intracranial hemorrhage, etc.), immunological findings (glomerulonephritis, Osler's nodules, rheumatoid factor), echocardiography suggestive of infective endocarditis, and microbiology (positive blood cultures that do not meet the main criteria).

THERAPY

After repeated blood samples have been drawn, a calculated initial therapy with broad-spectrum antibiotics must be started. Once the exact antibiogram is determined, the antibiotics must be changed as appropriate. A surgical valve replacement may be considered depending on the valvular damage present.

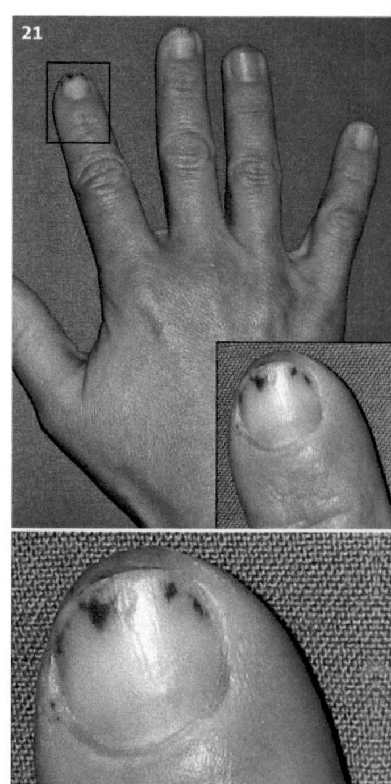

21 Clinical picture of splinter haemorrhages

SECONDARY HYPERTROPHIC OSTEOARTHROPATHY

EPIDEMIOLOGY

Little is known about the epidemiology. One study showed that signs of hypertrophic osteoarthropathy can be found in 4.5% of lung cancer patients.

ETIOLOGY AND PATHOGENESIS

Hypertrophic osteoarthropathy is associated with lung cancer, pulmonary infections, cystic fibrosis, right-left cardiac shunts and rarely with other diseases (e.g., Hodgkin's lymphoma or cirrhosis).

It is believed that the localized activations of endothelial platelet cells, with the subsequent release of fibroblast growth factors, play a major role in the pathogenesis.

One hypothesis for why lung diseases are associated with Morbus Pierre-Marie-Bamberger is that the normal megakaryocytes escape from fragmentation into platelets in the lungs. These megakaryocytes reach the distal extremities, where they release growth factors. Another hypothesis describes the direct production and release of growth factors by lung tumors.

HAND MANIFESTATIONS

There are often no symptoms, and it is possible that stress pain occurs. Pachydermia (abnormal thickening of the skin), periostosis, swollen fingers, seborrhea and hyperhidrosis clinically impact the hand. Watch-crystal nails (convex curvature in the longitudinal and transverse directions) are characteristic.

FURTHER CLINICAL MANIFESTATIONS

The other clinical manifestations predominate and are caused by the symptoms of the associated disease.

DIAGNOSIS

X-ray imaging shows pathognomonic changes of exclusively diaphyseal periosteal appositions in the long bones as well as in metacarpal bones, proximal and middle phalanges. The changes are usually symmetrical. If signs of hypertrophic osteoarthropathy are detected, the thorax should be examined to exclude localized pathological processes.

THERAPY

NSAIDs may alleviate the symptoms, but the causative disease must be treated.

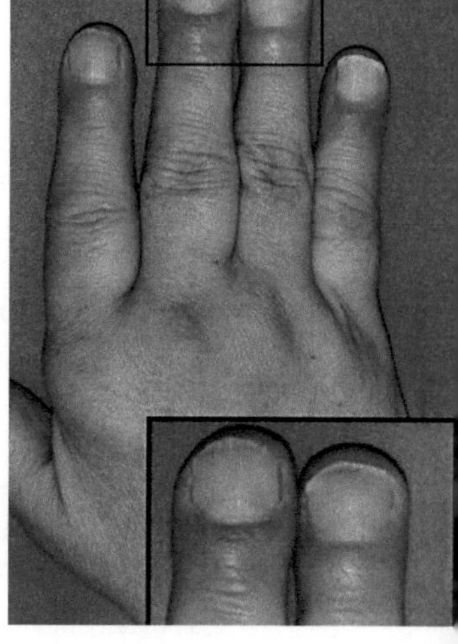

22 Clinical picture with watch-glass nails and thickened phalanges

CHRONIC REGIONAL PAIN SYNDROM (CRPS, MORBUS SUDECK)

EPIDEMIOLOGY

The frequency data in the literature vary between 2% and almost 40%. Chronic regional pain syndrome (CRPS) is 3 times more common in women than in men. The peak incidence is between the ages of 50 and 70; however children may already be affected.

ETIOLOGY AND PATHOGENESIS

The exact etiology is not known. In addition to the main triggers of the syndrome (operations, a pollutant or a trauma to the affected extremity), other painful events (e.g. heart attack, brain tumors, spinal cord lesions) are described as triggers. The intensity of the trigger event does not correlate with the severity of CRPS. The picture of CRPS is caused by varying degrees of interaction among different components: dysregulation of the sympathetic nervous system (micro-circulation disorders with hypoxia and metabolic acidosis), a neurogenically induced inflammatory process, a genetic predisposition, changes in the CNS, and immobilization or dysfunction of the limbs.

HAND MANIFESTATIONS

CRPS is manifested as a triad of sensory, autonomic and motor symptoms. The sensory components display a spontaneous pain that has a burning, cramping or gnawing quality. The pain is provoked by the slightest movements, temperature changes or contacts. Autonomous findings include soft tissue swelling and temperature fluctuations of the affected region such that the skin can be both cold and warm. Eccrine sweat gland activity may be disturbed in the forms of hyper- or hypohidrosis. The result over time is an active and passive functional limitation. Tremor, dystonia and lack of coordination can be observed. Trophic disturbances are late-stage signs; skin, hair and nail growth are disturbed, and the bones can be osteoporotically altered by the trophic disorder as well as by the immobilization.

FURTHER CLINICAL MANIFESTATIONS

CRPS is by far most commonly encountered in the hand; therefore, the clinical manifestations are mainly limited to this site.

DIAGNOSIS

The diagnosis is made according to history and clinical examination. On radiographs, osteopenia can be visualized, and bone scans demonstrate a diffuse enhancement of the affected site. Thermography reveals a temperature difference between the affected and the unaffected limb that can be measured, but this is not necessary in clinical practice. Other helpful tests are sudometry (sweat intensity measurements) or diagnostic sympathetic blockade (pain relief by drug inhibition of the autonomic nervous system).

THERAPY

The affected limb must be moved, touched and stimulated despite the pain. Medications such as NSAIDs, steroids, antidepressants, anticonvulsants, opioids, etc., may relieve pain in some patients. Interventions such as a sympathetic block, sympathectomy, epidural clonidine, intrathecal baclofen or acupuncture may lead to improvement. CRPS may also be treated with psychotherapy and occupational therapy.

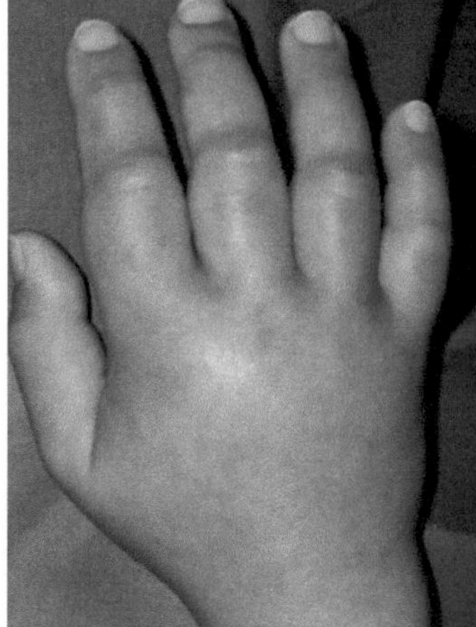

23 Clinical picture of soft tissue swelling around the hand

REFERENCES

Micheroli R, Kyburz D, Ciurea A, Tamborrini G,
The rheumatic hand – diagnostics in practice, Praxis
(Bern 1994). 2011 Sep 7;100(18):1097-106.

OSTEOARTHRITIS (OA)
→ Theiler R, Arthrose, Praxis Schweiz Med Forum
2002; 23:555-561
→ Häuselmann HJ, Hedbom E, Von der Pathogenese
der Arthrose zu therapeutischen Empfehlungen und
Knorpelersatz, Praxis Schweiz Med Forum
2002; 25:610-625
→ Zhang W, Doherty M, Leeb BF, et al. EULAR evidence-
based recommandations fort the diagnosis of
hand osteoarthritis: report of a task force of ESCISIT,
Ann Rheum Dis 2009; 68:8-17

RHEUMATOID ARTHRITIS (RA)
→ Tutuncu Z, Kavanaugh A. Rheumatic Disease in
the Elderly: rheumatoid Arthritis. Clin Geriatr Med
2005; 21:513-525
→ Rooney J. Oh, those aching joints! Helping Patients to
manage arthritis. LPN 2006; 2:26-34
→ Tamborrini G, Backhaus M, Schmidt W, et al. Ultraschall
und Arthritis, Z Rheumatol 2010; 69:889–902

GOUT (URATE ARTHROPATHY)
→ Baker JF, Schumacher HR, Update on gout and
hyperuricemia, the international journal of clinical
practice 2010; 64:371-377
→ Richette P, Bardin T, Gout, Lancet 2010; 375:318-328

**CALCIUM PYROPHOSPHATE
DIHYDRATE DEPOSITION DISEASE (CPPD)**
→ Announ N, Guerne PA, Diagnose und Therapie der
Kalziumpyrophosphatkristall-inuzierten Arthropathie,
Z Rheumatol 2007; 66:573-578
→ Weber U, Kalziumpyrophosphat-Kristallarthropathie,
Schweiz Med Forum 2009; 39:914-921
→ Tausche AK, Winzer M, Aringer M, Chondrokalzinose
und Pseudogicht – aktueller denn je, Akt Rheumatol
2008; 33:275-280

→ Hirose J, Clinical presentation and diagnosis of calcium deposition disease, Int. J. Clin. Rheumatol. 2010; 5:117-128

PSORIATIC ARTHRITIS (PSA)

→ Taylor W, Gladman D, Helliwell P, et al. Classification Criteria for Psoriatic Arthritis, Arthritis and Rheumatism 2006; 54:2665-2673

→ Gladman D, Clinical, radiological, and functional assessment in psoriatic arthritis: is it different from other inflammatory joint diseases? Ann Rheum Dis 2006; 65:22-24

REACTIVE ARTHRITIS (REA)

→ Frey D, Tyndall A, reaktive Arthritiden – was gibt es Neues? Schweiz Med Forum 2002; 24:577-580

→ Kim PS, Klausmeister TL, Orr DP, Reactive Arthritis: A Review, Journal of Adolescent Health 2009; 44:309-315

→ Petersel DL, Sigal LH, Reactive Arthritis, Infect Dis Clin N 2005; 19:863-883

SYSTEMIC SCLEROSIS (SSC)

→ Tamborrini G, Distler M, Distler O, Systemische Sklerose/Sklerodermie, MMP 2008; 31:1-9

→ Erre GL, Marongiu, Fenu P, et al. The "sclerodermic hand": A radiological and clinical study, Joint Bone Spine 2008; 75:426-431

→ Distler O, Gay S, Sklerodermie, Internist 2010; 51:30-38

DERMATOMYOSITIS/POLYMYOSITIS (DM/PM)

→ Stoltenburg-Didinger G, Genth E, Dermatomyositis, Zeitschrift für Rheumatologie 2009:1-8

→ Kirrstetter M, Distler O, Polymyositis und Dermatomyositis – Manifestationsformen und aktuelle Behandlungsmöglichkeiten, Rheuma Nachrichten 2009; 51:10-13

SYSTEMIC LUPUS ERYTHEMATOSUS (SLE)

→ Göhner K, Distler O, Manifestationsformen des systemischen Lupus erythematodes und aktuelle Behandlungsmöglichkeiten, Rheuma Nachrichten 2009; 50:8-13

→ Bleifeld CJ, Inglis AE, The Hand in Systemic Lupus Erythematosus, J Bone Joint Surg Am. 1974; 56:1204-1215

→ Rahman A, Isenberg DA, Systemic Lupus Erythematodes, N Engl J Med 2008; 358:929-939

DIABETIC CHEIROARTHROPATHIA (DIABETIC HAND-SYNDROM)

→ Philips P, Diabetes and the skin, Australien Family Physician 2005; 34:1037-1038

→ Kashyap AS, Anand KP, Kashyap S, et al. Diabetic Cheiroarthropathy, Postgrad Med J 2009; 85:43

ENDOCARDITIS

→ Silverman ME, Upshaw CB, Extracardiac Manifestations of Infective Endocarditis and Their Historical Description, Am j Cardiol 2007; 100:1801-180

SECONDARY HYPERTROPHIC OSTEOARTHROPATHY (MORBUS PIERRE-MARIE-BAMBERGER)

→ Guarneri C, Guarneri F, Vaccaro M, et al. PIERRE-MARIE-BAMBERGER SYNDROME, JAGS 2008; 56:184-185

→ Cannavo SP, guarneri C, Borgia F, Vaccaro M, Pierre Marie-Bamberger syndrome (secondary hypertrophic osteoarthropathy), International Journal of dermatology 2005; 44:41-42

→ Jajic I, Nails "Obstructing" Finger Growth in Length in Primary Hypertrophic Osteoarthropathy, Acta clin Croat 2000; 39:191-192

CHRONIC REGIONAL PAIN SYNDROM (CRPS, MORBUS SUDECK)

→ Köck FX, Borisch N, Koester B, Grifka J, Das komplexe regionale Schmerzsyndrom Typ 1 (CRPS1), Orthopädie 2003; 32:418-431

→ Troeger H, Hasenböhler P, Geschichte, Stadieneinteilung und Klinik des komplexen regionalen Schmerzsyndroms (CRPS), Handchir